Published by Creative Education
123 South Broad Street, Mankato, Minnesota 56001
Creative Education is an imprint of The Creative Company

Designed by Stephanie Blumenthal

Photographs by Archive Photos, Frank Balthis, Gary Benson,
Lisa Christenson, Richard Cummins, Victor Englebert,
Bonnie Sue Rauch, Mrs. Kevin Scheibel, Karlene Schwartz,
Tom Stack & Associates (Brian Parker, Inga Spence),
Unicorn Stock Photos (Doug Adams, Eric Berndt, Wayne Floyd,
Jeff Greenberg, Ed Harp, Andre Jenny, Tom McCarthy, Rob &
Ann Simpson, Aneal Vohra), The Viesti Collection, Ingrid Wood

Library of Congress Cataloging-in-Publication Data

Peterson, Sheryl.
Sugar / by Sheryl Peterson.
p. cm. — (Let's investigate)
Includes index.
ISBN 1-58341-190-9
1. Sugar—Juvenile literature. [1. Sugar.] I. Title. II. Series.
TP378.2 .P48   2001
664'.1—dc21   00-047386

First edition

2 4 6 8 9 7 5 3 1

SUGAR

SHERYL PETERSON

Creative Education

SUGAR

ORIGIN

The word "sugar" comes from the ancient Sanskrit word sharkara, *which means "material in a granulated (or a small, grain-like) form."*

4

Right, naturally sweet apples
Below, coffee sweetened with sugar

Most people love sugar. They sprinkle it on food, stir it into hot and cold drinks, and nibble it in the form of candy. The word "sugar" usually brings to mind images of sugar cubes and spoonfuls of refined white crystals. But sugar can take many different forms. And it can be found in some very unexpected places.

SUGAR
PLANTS

The total quantity of all sugars formed each year by all the plants on the planet has been estimated at 400 billion tons (363 billion t).

Sugar is found in all fruits and vegetables. Green plants take in the sun's rays and manufacture sugar. Sugar gives heat and energy to animals, including humans, faster than any other **nutrient**. It provides power for the body much as gasoline provides power for a car.

Above, young sugarcane leaves
Right, rows of sugarcane plants

SUGAR HISTORY

The story of sugar's discovery and rise to importance is woven into all cultures. Early peoples did not use sugar as we do today—as something to add to food to make it taste better. Their survival depended, in part, on finding sweet fruits (which gave them calories and vitamins) and avoiding bitter plants (which were often poisonous).

SUGAR
USAGE

Early European chemists mixed sugar into their medicines. The sweetness helped to mask the bitter taste.

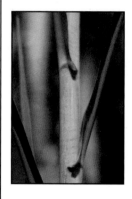

Above, detail of a sugarcane stalk

SUGAR

8

SUGAR

The first **refined** sugar, called **sucrose**, came from sugarcane, but its geographical origins are a mystery. It came from either India or the South Pacific. From there, word of this exciting "spice," as it was then called, traveled to the Middle East, China, and Egypt. Some time later, Arabian armies carried sugar plants across the coast of Africa and up to Spain. By 600 A.D., the practice of breaking up sugarcane stalks and boiling them to produce sugar **crystals** was widely known. The medieval European world was quick to appreciate what sugar could do for food, and a flourishing trade developed.

A farmer carrying cut sugarcane

Soldiers in Persia saw sugarcane growing on the banks of the River Indus and called it "reeds which make honey without bees."

9

On his second voyage to the New World in 1493, Christopher Columbus brought cane cuttings to Hispaniola, which is now Haiti and the Dominican Republic. Columbus and his men may have traded this sweet crop to the natives for gold, gems, and spices.

Above, boys eating raw sugarcane
Left, a sugarcane farm

SUGAR
SAYING

In England, during the reign of Elizabeth I (1558–1603), sugar was still rare enough to be a luxury. Only the rich could afford it, and they loved it so much that they added it to wine, fruit, and practically any other food or drink. Some of England's wealthiest are even said to have blended it with crushed pearls and gold to make medicines. In the 16th century, a teaspoon of sugar cost the equivalent of a U.S. five-dollar bill in the year 2000.

A bowlful of sugar cubes

At that time, poorer people generally used honey or **molasses**, a by-product of cane sugar, to sweeten their food. But both of these sweeteners added a strong flavor of their own. Soon the common people demanded sugar. Merchants filled huge ships with sugarcane from places as far away as the Canary Islands to bring back to Europe.

SUGAR
G I F T

Sugar shaped into a small loaf was a very special and expensive gift in early times.

Left, raw honey with honeycomb
Below, a bottle of processed honey

SUGAR
USAGE

Around the world, a single person eats an average of 64 pounds (29 kg) of sugar a year.

12

With increased demand, the sugar trade expanded, and so did pirating and slave trading. Pirates hijacked ships to seize their cargoes, and sugar was a fine prize. The slave trade, which continued for two centuries, grew in part to supply sugar plantations with workers. Although sugar wasn't the only crop grown by slave labor, it was one of the major ones. Slaves were captured and transported on ships from Africa to Brazil, the Caribbean, and North America. The strongest slaves were sent to cut sugarcane and haul it from the fields. The work was hot and difficult, and large numbers of slaves died while harvesting cane.

An illustration of slaves harvesting sugarcane

MAPLE SUGAR

As time passed, sugarcane pro-duction grew to include more countries, especially those with hot, humid climates. Another sugar, however, had been made for centuries in North America near the Great Lakes and the Saint Lawrence River. Native Americans made sugar from the sap of the sugar maple tree.

A sugar maple tree can live 300 to 400 years. There are 12 species of sugar maples, but the Acer saccharum (A-sir SACK-uh-rum) pro-duces the greatest amount of sugar.

13

Above, a maple sap collection bucket
Left, a sugar maple tree in fall

SUGAR
MAKING

Early maple sugar makers thought their sugar was better than cane sugar since it was made during cold seasons. They argued that insects could feed on and get mixed into the sugar grown in hot climates.

14

A legend from the Ojibway tribe says that the first sugar maker was the common red squirrel. The hungry squirrel sought out the thin-skinned maple branches in late winter and nipped and bit until sap oozed out. Then the squirrel ate the sugary crust that formed on the tree bark. Eventually, Native Americans became curious and tried the sugar trees themselves.

A sugarhouse and sap collection buckets

When spring came, Ojibway families headed for the **sugar bush** to **tap** maple trees. They cut a V-shaped slash into the maple tree trunk, inserted a stiff piece of bark for a trough, and let the sap flow into a hollowed-out log. Then they collected the sap and boiled it. The flavor became stronger as the sap boiled.

SUGAR

The universal rule for making maple sugar is, "Never tap a tree that is smaller than the bucket you're putting to it."

15

SUGAR
WEATHER

Native Americans tapped sugar maples when the weather was between 35° and 45° F (1° to 7° C) during the day and below freezing temperatures at night.

In time, European settlers learned how to make what the Native Americans called "sweet water." In colonial North America, sugar-on-snow parties were held to celebrate the end of winter and the arrival of spring. As a special treat for the children, hot maple syrup was poured on clean snow to make a sticky taffy called jack-wax. Since it was hard for the settlers to keep syrup fresh, most of it was cooled into small maple sugar cakes.

An illustration of Native Americans tapping maple trees

The sugar maple leaf is Canada's national symbol. The large red leaf can be found on the Canadian flag.

17

Today, maple syrup is boiled in a sugarhouse, and new equipment, such as **spiles**, metal buckets, tubing, and plastic bags, speeds up the process. Modern maple sugar makers help keep alive an important time in history, when sugaring was a way of life. Many people still use maple syrup because they like its natural ingredients and delicate taste.

Above, the Canadian flag Left, maple sap dripping from a spile

SUGAR
LEADERS

Brazil leads the world in cane sugar production, and France leads in beet sugar production.

SUGAR PRODUCTION

In 18th-century France and Germany, researchers were looking for a way to grow sugar in colder climates. They found a plant called the sugar beet, which had grown wild in ancient Greece and Asia. By the middle of the 20th century, sugar beet farms could be found in Russia, the United States, and Canada. More and more people were eating sugar, and the demand for it continued to grow.

Harvested sugar beets

Today, sugarcane growers and sugar beet growers compete with each other around the world. Cane provides 56 percent of the world's sugar, and beets provide 44 percent.

19

Below, a healthy golden sugar beet

SUGAR

A substance called bagasse (buh-GAS) remains after sugarcane is processed. It can be burned as fuel, made into paper, or used as livestock feed or chicken litter.

20

Above, a sugarcane field on fire
Right, a man cutting full-grown sugarcane stalks

SUGARCANE

Sugarcane is a giant **perennial** grass that thrives in moist climates such as those of Brazil, Hawaii, and Cuba. It grows to be 8 to 24 feet (2.5–7 m) high and stores sugar in its stalk. To grow, sugarcane needs 10 to 12 months of warm weather and a large amount of water. Before cane fields are harvested, they are set on fire for a few minutes to burn off the dry leaves. Then workers with long, heavy knives called machetes chop down the stalks. Some of the work is done by machine, but most must be done by hand.

21

I n mills near the cane fields, raw sugar is separated from the plant. The stalks are washed and cut into shreds by rotating knives, and huge rollers press out the juice. The juice is then boiled to a clear brown syrup and crystallized in a machine that spins like a washing machine.

Transferring boiling cane juice from one vat to another

Before it was packaged in bags, sugar was scooped out of large wooden barrels for customers.

Above, a girl carrying sugarcane stalks
Right, raw sugar

Finally, the raw sugar is taken to a refinery by ship or railroad car. There, **centrifuges** spin the raw sugar with water to separate the molasses from the crystals. When the sugar crystals are washed and dried, they have become table sugar.

Sugar beets grow in cooler regions of the world. The beets are planted early in the spring and take five months to grow. The root of the plant, where the sugar is made, is cream-colored and about one foot (30.5 cm) long. Eight-inch (20 cm) stems rise from the center of the root and bear big, dark leaves.

SUGAR
WEIGHT

Each sugar beet weighs about two pounds (907 g) and contains about 30 teaspoons (148 ml) of sugar.

23

The cream-colored root of a sugar beet

SUGAR
GLASS

Sugar in a solid form is used to make the harmless stunt glass that is seen shattering in action movies.

Like all food crops, sugar beets need both sunny weather and rain to produce a healthy yield. Too many cloudy days or too little moisture will slow their growth. Beet fields are kept free of weeds and harmful insects with tilling and careful pesticide applications.

Harvesting sugar beets with machines

In the fall, when the evenings turn cool, the beets begin storing sugar in their roots. Growers leave the beets in the ground until the sugar content reaches its highest level. Then the growers have to harvest them quickly, before the sugar level drops. Farmers work day and night, driving machines that cut off the beet tops and dig up the beets. Then they truck them to a sugar factory for processing.

25

Above, a spoonful of refined white sugar

SUGAR

*The **taste buds** that recognize sweetness are at the very tip of a person's tongue.*

Right, a huge mound of sugar beets
Below, the sugar from beets is used in pies and other desserts

At the factory, workers wash the beets and use a machine to chop them into thin strips. These pieces are put into enormous containers of hot water called diffusers, which change the sugar to syrup. The leftover pulp is made into feed for farm animals.

SUGAR
DIET

One teaspoon (5 ml) of sugar contains 16 calories, no fat, and no sodium.

The syrup is then boiled down to a thick brown mush. This mixture is filtered, purified, and dried until it becomes tiny white crystals. Most beet sugar, like most cane sugar, is packaged in bags, but some of it is pressed into sugar cubes for tea and coffee drinkers.

Boxed sugar cubes

SUGAR

FLAVOR

Many so-called sug-arless foods contain unrefined sweeten-ers, such as honey, molasses, raisins, dates, or fruit juice.

28

SUGAR IN FOOD

Sugar adds to the flavor, aroma, color, and texture of food. In baked goods, such as bread, sugar is food for the yeast. Soft drinks use a corn syrup sugar, which supplies the drink with body and makes the drinker feel full. Sugar gives candy a smooth texture. In jams and jellies, sugar helps hold the fruit together and keeps mold and yeast from growing. Ketchup relies on sugar to blend its flavors and soften the acidic taste of the tomatoes. Molasses, once the main sweetener available to people who weren't wealthy, is still used in baking, too.

Sugar-sweetened peanut butter cookies

SUGAR
SYRUP

Forty gallons (151 l) of maple sap can be converted into one gallon (3.8 l) of maple syrup.

Sugar is usually thought of as food, but it is also used in non-food products, including cement, plastics, nylon, and medicines. Mirrors, printing inks, and some pesticides are made with sugar, too.

Many medicinal tablets and capsules include sugar

SUGAR
TYPES

Brown sugar, also called soft sugar, is made of sugar crystals that are coated in molasses syrup. Powdered sugar is granulated sugar that has been ground to a smooth powder and sifted.

SUGAR AND HEALTH

Scientists do not always agree on how refined sugar affects people's health. Some researchers argue that people should avoid refined sugar. Others say that adding sugar to healthy foods makes them taste better, so people will eat more of them.

Scientists do agree that active people need a daily supply of the natural sugars found in fruits, vegetables, and grains. They agree that people who eat too much refined sugar may store the sugar as fat and become overweight. They also agree that sugar is one of the main causes of dental cavities.

Years ago, sugar was in short supply and considered very valuable

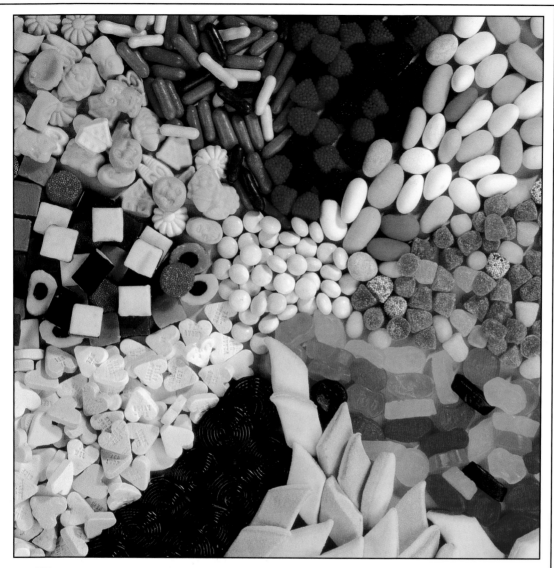

Some people's bodies are unable to process sugar normally. This disease is called diabetes. It is incurable but can be controlled with medicine and changes in diet.

Above, a gingerbread man cookie
Left, sugared candies and sweet treats

Sugar is no longer a luxury of the rich. The sugar bowl has become a standard fixture in many homes around the world. From its earliest beginnings to modern times, sugar production has continued to grow to keep up with the human craving for sweets.

Glossary

Machines with rotating containers used to separate molasses from raw sugar are called **centrifuges**.

Crystals are small, clear pieces of sugar.

Molasses is a thick, dark syrup produced during the process of refining sugar.

A **nutrient** is a vitamin, mineral, protein, or carbohydrate in food that contributes to good health.

A **perennial** is a plant with a life cycle of more than two years.

Once cane and beet sugar have been washed, filtered, and processed, they are called **refined** sugar.

Spiles are spouts that are inserted into sugar maple trees to act as troughs.

The chemical name for common table sugar is **sucrose**.

Sugar bush is the name given to a grove of sugar maple trees.

To **tap** is to drill a small hole in a sugar maple trunk, allowing sap to run out.

Taste buds are tiny organs on the tongue that identify a food as sweet, salty, sour, or bitter.

Index